GLOBAL WARMING

What Else Can You Do About It?

Elton Eastwood MD

iUniverse, Inc.
Bloomington

Global Warming
What Else Can You Do About It?

iUniverse books may be ordered through booksellers or by contacting:

iUniverse
1663 Liberty Drive
Bloomington, IN 47403
www.iuniverse.com
1-800-Authors (1-800-288-4677)

ISBN: 978-1-4620-1150-6 (pbk)
ISBN: 978-1-4620-1153-7 (ebk)

Printed in the United States of America

iUniverse rev. date: 06/02/2011

Chapter 1

Is the Earth Really Getting Warmer

When we get temperatures as low as 20 degrees below zero, and have to walk through a foot of snow, we tend to say, "So much for global warming," Maybe we could use some global warming around here. But the process does continue. Green house gases such as carbon dioxide [CO_2], from burning fossil fuels, continue to accumulate in a layer up in the atmosphere, and continue to warm the earth and melt the ice caps.

Scientists have taken ice cores in Greenland and the Antarctic, which contain air bubbles, and have checked the CO_2 levels of the air in the bubbles and found it to be 200 parts per million, [ppm] most of the time, but every 100,000 years the CO 2 level would rise and reach a peak of 300 ppm. The ice cores are like the rings in a tree stump because there are clear ice strips alternating with the winter ice that is made from compacted snow and is more opaque and contains the air bubbles. These rings can be counted back 400,000 years.

They can also determine the temperature at any level. The peaks of CO_2 occur when the temperatures peak but they don't go above 300. Then in about 1950, there was an increase in the rate of global warming and now the CO_2 level is at 380 ppm. This increase in the rate of warming is attributed to the green house gases from burning fossil fuels. Eleven of the past twelve years are the warmest since reliable records began around 1850. The odds against such warm years happening in sequence by chance are extremely great. The earth temperature was 0.74 degrees Celsius, (1.32 degrees F.) higher in 2005, than in 1956.This rise in temperature has been attributed to the increased effect of green house gases such as methane, nitrous oxide, and especially carbon dioxide.

Before the industrial revolution, Earth's atmosphere contained about 280 parts per million of CO_2. In the late 1950s, the level of CO_2 had reached 315 ppm. Now it is 380 ppm and has been increasing at roughly 2 ppm annually. Methane levels are about two and a half times pre-industrial levels, and nitrous oxide is about 20% higher. CO_2 levels have taken an alarming jump since the year 2000. An international team of scientists reported that CO_2 emissions have increased from 1.1 per cent a year in the '90s to 3.1 per cent per year since 2000. If they continue to increase at this rate, they would rise to 560 ppm in 2050 and 1390 ppm by 2100.

Scientists say we should not exceed 450 ppm. If it goes beyond that, the Greenland and Arctic ice sheets will melt and the sea levels will rise. CO_2 levels vary in cities. It is 450 in Baltimore, 550 in Phoenix, and 700 on a bad day in New York City. U.S. carbon emissions make up one fourth of the world total. This year China is expected to surpass the U.S.

The U.N. Intergovernmental Panel on Climate Change has listed the following as current effects of global warming: Warmer water is killing coral reefs. Tropical frogs are dying. Drought is a problem in Africa and it will continue to get worse with global warming.

Molecules of CO_2 stay in the atmosphere as long as 200 years. CO_2 from the tail pipes of Model T Fords are still up there. Reductions in CO_2 as called for in the Kyoto agreement will only slow the continued accumulation. It therefore becomes necessary to devise a way of removing significant amounts of CO_2 from the atmosphere.

Scientists say that CO_2 emissions must be cut by 50% in the next 50 years in order to avert a future global warming disaster. Rajendra Pauchari, head of the Intergovernmental Panel on Climate Change, says that in the 20th century, sea level rose about 17 cm. [6.8 inches]. Predictions by the end of this century are for up to 59 cm.[23.6 inches]. The IPCC also projected loss of sea ice in the Arctic and Greenland to be 2.5% per decade from 1953 to 2006. It has actually been 7.8% or 30 years ahead of projections.

Since 1993, satellite observations of changes in sea level show it is rising at a rate of 3.1 mm per year. Photos of Arctic sea ice, (See Frontspiece) in 1979 and 2005, show a marked decrease. Estimates for rise in sea level in the 21st century are from 30 to 40 cm. By the 2080s, the sea will flood the homes of millions of people, especially in the large deltas of Asia and Africa. This would put about 25000 square miles of coast-line under water in the lower 48 states, an area equivalent to the state of West Virginia. Louisiana, Florida, North Carolina, and South Carolina. Texas would lose the most land. New Orleans would be reduced to a narrow piece of land along the Mississippi River, leaving the French Quarter and the oldest neighborhoods. Water would cover the first few blocks of Bourbon St. In Boston, land along the Charles River would be submerged, along with the band shell where the Boston Pops Orchestra plays. Some runways at Logan Airport would be under water.

In New York City, at the southern tip of Manhattan, Battery Park City, home to 9000 people would be inundated. Mansions of the Hamptons would be swamped by the surf. Runways at La Guardia Airport would be partly under water. A medium size storm surge could wipe out thousands of homes in low parts of Brooklyn and Long Island.

In Miami, South Beach, the Everglades, and the airport would be under water. Many of the beach resorts would be gone. Spits of land would be left in South Beach and Fisher Island.

Fisherman's wharf in San Francisco would be submerged. So would the Giants' baseball park, Silicon Valley and the San Juaquin and Sacramento deltas. Also under water would be the Embarcadero Waterfront, runways at the San Francisco and Oakland Airport and the new Oakland As stadium planned in Fremont.

How do you survive the rising sea level? You can move inland but at a cost of trillions of dollars. You could protect yourself with barriers, sluices, pumps etc. but again the cost would be prohibitive .Buildings could be raised on stilts, or built so that they will float as the water rises as they are now doing in Holland. Perhaps it would be cheaper to curtail green house gases.

Scientists have found that large tracts of sea ice are now only 3 feet deep—half the thickness of 6 years ago. They used a torpedo shaped device towed by a helicopter that beamed electromagnetic waves at the ice surface. Satellite images show that Arctic sea ice is at its smallest size ever since measurement began 30 years ago.

There is now about 1.15 million square miles of sea ice in the Arctic. This is about 38,600 square miles less than in 2006. Average loss of sea ice has been about 38,600 square miles each year in the past decade. Melting of Arctic ice accelerates global warming. In one to two decades their will be no sea ice in midsummer in the Arctic, scientists say. This will decimate the polar bears. The penguins will not be happy about any attempt to transplant them to the Antarctic. They also say it will cause a severe drought in southwest U.S.

I need to emphasize that most recent studies show Arctic ice is disappearing at an alarming rate, much faster than had been predicted. Antarctic ice has also been melting. Probably the only good thing that could come out of establishment of a Northwest Passage is that it would shorten the trip from London to Tokyo by 3000 miles when compared with going through the Suez Canal.

CHAPTER 2

Are Humans the Cause of Global Warming?

Some people say that this is just a normal cycle of warming. The earth has been warm before. A few years ago, scientists were warning that we might be going into another ice age. These people admit that there may be warming, but it's not our fault. The IPCC in 2009, concluded with 90% confidence, that humans are causing global warming.

Coal provides 50% of American power. Power generation is the biggest source of CO_2 in America. 40% of CO_2 emissions come from coal in the U.S., which has been the world's largest consumer of coal. In the future, 85% of CO_2 is going to come from developing countries, primarily China and India. China has plans to build 650 coal fired plants before 2012, and the emissions from those will be quite substantial.

World green house gas emissions now come from:

Electricity and heat 24.5%
Deforestation 18.2%
Industry 13.8%
Transport 13.5%
Agriculture 13.5%
Waste 3.6%
Other 12.9%

It has been said that to accommodate the aspirations of more than 5 billion people in the developing countries, the size of the world economy should increase by a factor of 4 to 6 by 2050. At the same time, global emissions of green house gases will have to remain steady or decline to prevent dangerous changes to the climate. After 2050, emissions will have to drop even further.

Green house gases must be cut by at least 60% over the next 50 years to avert the worst consequences of global warming. The McKinsey Global Institute projects that the number of China's vehicles will increase from 26 million in 2003, to 120 million in 2020. 1400 new cars hit the road each day. The US has 230 million cars. In China,average residential floor space will increase 50%, and energy demand will grow 4.4% annually.. For coal fueled plants, carbon capture and sequestration under ground is necessary, but much expensive research must be done.

If the world wants a solution, we must develop ways to capture CO_2 from the air. Driving more hybrid cars and using spiral fluorescent light bulbs will not solve the problem. That doesn't mean that we shouldn't do them. Every thing helps. Oil consumption in the U.S. per person is 26 barrels a year, compared to 12 for Europeans. We have 5% of the world's population and consume 25% of the world's oil. We have 3%of the world's oil reserves. World wide demand for oil is growing by 2 to 3 million barrels a day every year.

In the United States, some 150 new coal fired power stations are on the drawing boards. In China, two 500 megawatt coal fired power plants are starting up every week. Standard pulverized coal generation can be made a bit cleaner by burning it at higher temperatures This can cut CO_2 emissions by a fifth.

Chinese leaders have staked the country's stability on a cheap currency and export surpluses, which create jobs for their people. They have a current surplus of 400 billion dollars for the year 2007.

Several years ago, we were concerned about the hole in the ozone layer. To correct this condition, The U.S. and Europe prohibited the use of fluorocarbons in spray cans. With the Kyoto agreement, those who signed on such as the European countries, issued permits to release CO_2 to their plants and factories. If they were going to have to produce more than the allowed amount, they would have to pay for it. This is the carbon trade, and it is called the cap and trade system. They could purchase and set up wind machines in developing countries as a better way of producing power, to obtain permission to increase emissions over that allowed by their permits.

By replacing CFC refrigerants with HCFC to close the ozone hole, a byproduct is produced called HCF23. That is a green house gas about 12000 times as strong as CO_2. The industrialized world has installed devices to remove this chemical. In China they have delayed the installation, and have received 2.7 billion dollars from European companies to install removal devices, which should have cost no more than 136 million. It is good to keep this stuff from getting into the atmosphere, and the companies got the right to produce CO_2 over the limits in their own factories. But China really made a profit.

The U.S. and China did not sign the Kyoto agreement. So U.S. plants have not had to deal with carbon permits and credits. In Europe there are two sources of carbon credits. The first is the allowance given by the government to the companies in the five dirty industries: electricity, oil, metals, building materials, and paper. The second source lies outside of Europe. It is linked to the clean development mechanism,CDM, set up under the Kyoto protocol.

This provides for emissions reductions in developing countries. If certtified by the U.N., they may be sold. 30 billion dollars were traded last year. In Germany, too many credits were awarded to coal fired power plants. The owners then charged customers for carbon costs that they did not have to pay. This happened in the Netherlands, Spain, and the U.K also. Prices for carbon fluctuated from $40 a ton down to $1 a ton. The $1 price resulted when it became clear that the government had over supplied with permits. Subsequently, the number of permits was reduced and the price came back up to $30 a ton.

CHAPTER 3

What Can Be Done About Dependence on Foreign Oil?

We are very dependent on foreign oil. We import 60% of our oil. One solution is the production and use of liquid coal. We have a lot of coal. One ton of coal produces only 2 barrels of fuel. When burned, 2 tons of CO_2 are produced. This is 4 to 8 per cent more than you would get from gasoline.

Law makers from coal states are proposing that U.S. tax payers guarantee billions of dollars in construction loans for production plants. It is estimated that it will cost 70 billion dollars to replace 10% of the gasoline.

The preferred method is to turn coal into gas before using it to generate electricity. The resulting CO_2 and hydrogen are then separated, the hydrogen used to generate electricity and the CO_2 is stored. These are integrated gasification combined cycle plants, IGCC. A few have been built. The cost is 60% higher than a pulverized coal plant. We should not build any more coal fired plants without IGCC. Carbon capture and storage, CCS, is being done in three places in the world, at Sleipner, Norway, in Salah, Algeria, and at the Weyburn oil field in Saskatchewan, where CO_2 from North Dakota is piped across the border and used to increase the pressure in a partly depleted oil field.

If 60% of the 1.5 billion tons of CO_2 that America produces every year, from coal fired power stations, were liquified for storage, it would take up the same amount of space as all the oil the country consumes. If it were stored as CO_2 gas,the space required would be much greater.

Chapter 4

Ethanol

There is much talk about renewable sources of energy. They now provide just 3.6% of the nation's energy, and the government predicts their share will grow to a grand total of 4.2% by 2030. By these calculations, it sure looks like a fossil fuel future for America.

World energy is produced by the following:

Coal 25%
Oil 34,3%
Natural Gas 20.9%
Nuclear 6.5%
Renewables 13.1%
Biomass 10.4%
Hydroelectric 2.2%
Other 0.5%

Other renewables are divided as follows:

Geothermal 0.41%
Solar 0.039%
Wind 0.064%
Tidal 0.0004%

The emphasis in the past few years has been on making ethanol. This will help to relieve the dependence on foreign oil, but will not do much about global warming. The production of ethanol is subsidized by the federal government. Congress called for production of 7.5 billion gallons of ethanol a year by 2012, up from about 4 billion gallons a year at the time the bill was considered. They will be burning more than that long before the deadline, thanks to government tax rules and subsidies.

According to the Renewable Fuels Association, ethanol production was more than 5 million gallons in 2006, compared to gasoline and diesel consumption of 140 billion gallons annually.

Corn ethanol is expensive. America's subsidy costs taxpayers between 5.5 billion and 7.3 billion dollars a year.

When you consider the emissions from tractors and farm equipment in raising the corn, ethanol is not so green, but some studies say there may be a 10% to 15% improvement over gasoline. Although CO_2 is emitted when it is burned, theoretically CO_2 is absorbed by the plants that are produced to make it.

Currently, ethanol accounts for only 3.5% of American fuel consumption. Any car can use 10% ethanol fuel, but only 6 million out of 237 million vehicles in America, can burn E-85, [85% ethanol]. Converting a vehicle costs about $200, but invalidates the guarantee. Detroit has promised that half of its cars will be able to use E-85 by 2012.

The energy derived from a gallon of ethanol is 67% of that obtained from gasoline. Bio-diesel gives 86% of regular diesel fuel. Producing corn ethanol consumes just about as much fossil fuel as the ethanol replaces. Bio-diesel from soy beans is only slightly better. About one fifth of the corn crop will be brewed into ethanol. Even if we turned all of our corn into fuel for cars and trucks, it would provide only 12% of our need for gasoline. The upper limit of corn based ethanol will be about 12 to 15 billion gallons.

We have to save some of the corn crop for livestock feed and for our own food. There are only a few hundred places that sell E-85. The supply chain is expanding slowly. A standard barrel of ethanol, 42 gallons, is worth about 28 gallons of gasoline because it contains only 80,000 British Thermal Units of energy compared to about 119,000 BTUs for unleaded regular gasoline. If you fill your tank with E-85, you will run out of gas about 33% sooner.

One problem is that ethanol requires large amounts of natural gas for its manufacture. Heating at various steps requires 36,000 BTUs of natural gas to produce a gallon of ethanol, with its 80,000 BTUs of energy. The U.S. will have to increase natural gas imports from outside of North America. Some plants are using coal which is not a good idea. Driving a mile on ethanol made with coal is worse than driving with gasoline as far as climate change is concerned.

Corn starch content is extracted to be made into ethanol, but there is still corn's protein and other nutrients. This becomes distiller's grains and can be fed to cattle. Ethanol from corn is not the ideal solution. It doesn't help much, if any, with the global warming problem. From 3% to5% of nitrogen used in fertilizer to raise crops could end up in the atmosphere as nitrous oxide, a potent long lived green house gas. Also it is said that there is an ever increasing dead zone in the Gulf of Mexico at the mouth of the Mississippi River from the effluent draining off the fields.

More ideal would be cellulosic ethanol, which, as years go by, may replace corn ethanol. This can be made from corn stalks, straw, switch grass, wood chips, or anything containing cellulose. President Bush has recommended that 385 million dollars be provided as a subsidy for its production. It may not be available for 5 years. It requires enzymes to free the sugar from the lignin. One company, IOGen, of Ottawa, Ontario, has a small plant, which uses "jungle rot", a fungus from Guam. Other investigators are using mushrooms to make their enzymes.

The government predicts a plant would cost five times as much as a corn plant, and would not be in operation until 2010. Range Fuels expects to get a plant working in 2008, which will use heat and pressure on timber industry waste and wood chips to change it to gas, then extract ethanol with a catalyst. With the help of government subsidy, their product can be competitive.

The company, BP, is developing a different fuel which is more energy intensive than ethanol. It is biobutanol.

Brazil uses sugar cane, which yields 800 gallons an acre, twice as much as corn. The cost of production is 87 cents a gallon. Ethanol provides 40% of Brazil's fuel for cars and light trucks.

Chapter 5

Nuclear Energy

Nuclear power provides 17% of the world's energy. Currently, 435 nuclear reactors are in operation throughout the world. Compared to energy from coal, or natural gas, they contribute very little to green house gases. Analysts predict a tripling of nuclear generation by 2050. Developing nations would account for about one third of this, so that in all countries, from 10% to 20% of electricity would be from nuclear plants.

Now, nuclear plants provide 18% in Britain, 19% in the U.S., and 80% in France. An application to build a plant can cost anywhere from 20 million to 100 million dollars. Cost of nuclear electricity is 6.7 cents per KWH, compared to 4.2 cents per KWH for pulverized coal.

There are other problems. There is only one steel plant in the world, in Japan, large enough to turn out a nuclear pressure vessel. About 25 reactors are under construction outside the U.S. China hopes to build 30 over the next 15 years. By 2030, the U.S. will need 60 to 70 more to replace the old units and build new ones.

In the U.S., there have been no new nuclear plants built for 29 years. NRG Energy, a Texas company, will request permission to build two new reactors, 85 miles south of Houston. The estimated cost is 5 to 6 billion dollars. At least 17 other companies have 29 plant proposals in the planning stage. There is a sudden rush to build because congress passed the 2005 energy bill, which included an estimated 15 billion dollars in subsidies to jump start the dormant nuclear industry.

Builders have to act early because only a few of the first reactors will get full benefits unless more money is appropriated. The cost to produce electricity in a nuclear plant is 65% higher per kilowatt than from coal, and nearly 6 times as expensive as that from natural gas.

Americans spend about 300 billion dollars per year for imported oil and gas, and an additional amount for military purposes. In 2006, the U.S. obtained 84.9 % of its energy from hydrocarbons. The US uses 21 million barrels of oil a day—27% from OPEC, 17% from Canada and Mexico, 16% from others, and 40% produced in the US.

Nuclear energy would be clean, but it has been discouraged by fear and claimed disadvantages, and dangers, that are actually negligable. The problem of nuclear waste has been politically created by the US government barriers to fuel breeding and reprocessing. Spent fuel can be recycled into new fuel. It doesn't need to be stored in expensive repositories.

Reactor accidents are brought up as possibilities, but there has never been even one human death associated with such an accident. All other forms of energy generation entail industrial deaths in mining, manufacture and transport.

At present, 43% of US energy consumption is used to produce electricity.

Arthur B. Robinson, Noah E. Robinson, and Willie Soon, of the Oregon Institute of Science and Medicine, suggested an environmentally sound way that the US can achieve energy independence. There are 104 nuclear power reactors in this country, with an average output in 2006 of 870 megawatts each, [a megawatt is 1 million watts] for a total of 90 gigawatts. [1 billion watts] If this were increased by 560 gigawatts, nuclear power could fill all current US electricity requirements, and have 230 gigawatts left over for export,or as hydrocarbon fuels replaced or manufactured.

Rather than a 300 billion dollar trade loss, the US would have a $200 billion trade surplus. Heat from the plants could be used for coal liquefaction or gasification.

This plan would call for nuclear sites in each of the 50 states with 10 reactors at each site. This plan is flexible and there could be more sites in California and two of the less populated states could share a site. It would cost a trillion dollars. At best, only 42% of the heat is utilized to produce electricity. The remainder could be used to produce 38000 barrels of liquified coal per reactor, giving us an output of 7 billion barrels of oil per year, with a value of 400 billion dollars per year. This is twice the oil production of Saudi

Arabia. They say the US has enough coal to sustain this production for 200 years. This would be a free market program, not a government program.

James Lovelock, the man who gave us the Gaia Hypothesis, namely, that the earth is an entity in itself, that attempts to regulate and control it's own destiny, is highly revered in the green movement. He has decided to come out for nuclear development as the best answer for global warming.

Jesse Ausubel endorses nuclear development also. He is head of the Program for the Human Environment at Rockefeller University. He says, "Measuring renewables in watts per sq. meter, nuclear has astronomical advantages over it's competitors." Others share his views.

According to the Department of Energy, in order to maintain nuclear's share of energy supply at 20%, the US needs to add 3 or 4 plants a year beginning in 2015. The US pumped nearly 7 billion tons of CO_2 into the atmosphere in 2005. More than 2 billion tons came from electricity generation. Fossil Fuels are burned to produce 70% of our electricity. About half of it comes from more than 500 coal fired plants.

America's need for electricity is expected to increase 50% by 2030. To meet the 2005 US electricity demand with wind, wind farms would have to cover 301,000 sq. miles., greater than the state of Texas. The amount of energy generated by one quart of the core of a nuclear reactor requires 2,5 acres of solar cells to generate that much power.

Public concerns about the safety of nuclear power have been fear of an accident with widespread fallout, and the hazards of nuclear waste. Since 9-11 a third worry is that it's a security risk.

A Chernobyl can't happen here. A survey by the Nuclear Regulatory Commission established that our reactors are free of the design flaws that permitted Chernobyl to explode. In the US, a typical reactor core is surrounded by multiple enclosures to block the escape of radioactive material, even if there was an accident.

In March, 2011 an earthquake that measured 9.0 on the Richter scale followed by a tsunami with an ocean wave 30 feet high, hit the northern Japan coast. There was shaking of a nuclear plant in Fukushima causing it to shut down. They lost electric power and they had a diesel or gasoline generator in the basement which was flooded by the wave and the generator would not work. They needed electricity to run the pumps that provided cold water to cool the waste fuel. The waste fuel boiled off the water in the tank and got hotter and hotter and hydrogen formed and there was an explosion. Some radioactive material did escape from the plant, and they had a difficult problem to handle.

China immediately canceled plans to build 45 new nuclear reactors. Others say we should not build any more plants in this country. Netanyahu, prime minister of Israel said they were planning for some nuclear plants but now he thinks they will stick with natural gas.

There is no question that nuclear has received a black eye from the Japanese experience, but we must take into consideration that most of our plants have a different system of cooling and would not have had the problem the Japanese had. We have 104 nuclear plants in this country that have been running without incident for 35 to 40 years except for the 3 mile island plant. They don't have to be built along the ocean. They can be built on higher ground. They should not be built near fault lines.

When you hear about a commercial airliner crashing killing all 300 passengers do you say we should not use airplanes any more or we should prohibit them from carrying more than 100 passengers? Let's be reasonable about this.

What about the waste? Spent fuel retains more than 95% of it's energy. It can be reprocessed and the volume of waste is reduced more than 60%. It is stored in concrete casks at the site of the nuclear plant.

NRG, an American company, is still waiting for a go ahead from NRC. They hope to be on- line in 2014 or 2015. The NRG reactor would be the latest version of reactors, though they have been used for 10 years. The water used to cool the core, and run the generating turbine, is also necessary to maintain the chain reaction. The neutrons that are emitted are traveling too fast to cause other atoms to fission. The water slows them down to allow the chain reaction to continue. In the event of an accident, cold water would flow to the core and the control rods would drop quickly, shutting down the reactor. This is reassuring, that the nuclear reactor can be brought under control so quickly.

The previous proposal for the expansion of nuclear fuel was that it should be recycled. Science writer Frank Von Hippel is opposed to the idea. He says the planned Yucca Mountain, Nevada depository for spent waste has not been opened, and is 2 decades behind schedule. At the earliest, it could be opened by 2017.

Now, nuclear waste is stored in steel and concrete casks, provided by the government, each containing 10 tons of waste. A 1000 megawatt reactor discharges enough fuel to fill two casks a year.

At present, if the spent fuel were stolen it would be difficult to make a bomb with it, because it is only 4 or 5 per cent U235 and the rest is U238. In the reactor, some of the U238 absorbs a neutron and becomes plutonium 239. It could be burned, but extraction and processing costs much more than the new fuel is worth. Recycling the plutonium reduces the waste volume only slightly. Moreover, it can be made into a bomb, if it gets stolen by terrorists, requiring additional security measures.

France has been a leader in reprocessing programs. The separated plutonium is combined with oxygen to make plutonium oxide, and that is mixed with U238 also as an oxide, to make a mixed oxide fuel, called MOX fuel. After this is used to make more power, there is still 70% as much plutonium as when it was manufactured. However,it is difficult to make a bomb out of it. It is stored back at the reprocessing plant.

Until recently, France, Russia, and the United Kingdom earned money by reprocessing the spent fuel of Japan and Germany. Anti nuclear activists in Japan and Germany, demanded that their government must show that they have a way of dealing with their spent fuel or close down the reactors. so, sending it abroad solved the problem temporarily. Their contracts specified that the spent fuel would eventually go back to the country of origin. So Japan has to find storage for the returning waste. They have been paying 1 million dollars per ton, about 10 times the cost of dry storage casks.

So France, Russia, and the UK have lost nearly all their foreign customers. The UK is planning to shut down their reprocessing plants, within the next few years, and it is estimated that it will cost 92 billion dollars to clean up the plant sites.

To summarize, Von Hippel says there are two reasons not to recycle fuel. First, that it is too expensive, and second, that terrorists might make a bomb with it. Separated plutonium is only weakly radioactive, and could be stolen and carried off by terrorists to make a bomb. Spent fuel on the other hand is highly radioactive and would require highly shielded containers to transport it.

The US nuclear industry concluded that "once through and storage" was the way to go. However, the Bush administration was attempting to revive the idea with use of new reactors that break down the recycled fuel so the volume of waste would be reduced.

The US National Academy of Science says that to recycle the first 62000 tons, the amount ready for storage in Yucca Mountain, would cost 50 billion to 100 billion dollars. They are thinking of using sodium cooled breeder reactors instead of water cooled, which could not be used for recycled fuel. Most countries have found them too costly to build and too troublesome to operate. Each sodium cooled reactor costs 1 to 2 billion more than the conventional reactor. It would take 45 to 75 of these to recycle the spent fuel from the 104 conventional reactors.

The cooling tanks at our nuclear plants are now filled to capacity with spent fuel, but there is plenty of room for more casks. We are expected to double the amount of our nuclear waste by the year 2050. We do need some place to store it.

The problem of nuclear waste deserves some discussion. Yucca Mountain in Nevada was supposed to be where it would eventually wind up in permanent storage. This area is already owned by the Federal government, and they have spent 22 years, and 9 billion dollars trying to get it ready for storage. The official opening was scheduled for 2017.

President Obama has submitted a budget, cutting funding for Yucca Mountain to a mere 197 million dollars, essentially dropping plans for Yucca Mt. storage. Some people say that, for the near term, this may be a good idea. At present, nuclear waste is kept in dry cask storage. More than 60,000 metric tons are in temporary storage at 131 sites around the country.

Nuclear fuel is in the form of a bundle of thin walled metal tubes, each filled with ceramic pellets of uranium oxide, the size of pencil erasers. It is extremely stable, and workers handle it with white gloves, not for the protection of the workers, but to protect the fuel.

It is lowered into a circular reactor vessel, and sealed up, and run for one to two years. Then the vessel is opened, and the oldest innermost fuel is removed, and the younger fuel is moved toward the center. Because of extreme heat, the fuel assemblies, when they are removed, are kept in water in a steel-lined concrete pool. Highly reactive fission products, such as strontium 90, and cesium 137 continue to burn. Heat production falls by 99% in the first year. After a few years, it has cooled down enough so that it can be removed from the water and placed in giant concrete casks near the reactor, where it is stored.

Uranium oxide, inside the reactor, splits and forms fission products which comprise about 3.4% of the waste. Some of the uranium picks up another neutron and becomes plutonium, about 1% of the waste. This is radioactive for more than 100,000 years. 95.6% of the spent fuel remains as uranium oxide. The recycling process separates out the uranium and plutonium and leaves the fission products.

There were plans to recycle the waste, but when President Ford was told that the process creates plutonium, and it could be stolen to produce a nuclear weapon, he banned recycling in 1976.

In 2009, President Obama's secretary of energy. Steven Chu announced that Yucca was now off the table. A one thousand megawatt reactor produces about 33 tons a year of spent fuel-enough to fill the bed of a large pick-up truck. It really isn't a large amount. Also, the longer it sits in a cask, the easier it is to handle as the heat diminishes in time.

Edward F. Sproat III, who was the official in charge of the Yucca project in the Bush administration, says all waste was destined to go to Yucca, and there is no other place to send it. He and others argue that President Obama and Senator Reid have the political power to block funding, but not to change the 1987 amendment to the Nuclear Waste Policy Act that targets Yucca exclusively.

Chapter 6

Geothermal Energy

Long running geothermal plants in the U.S. generate more electricity than all of the solar and wind installations. The U.S. is the world's biggest producer of geothermal energy, with plants in the western hot springs and geyser regions, but they are all developed on sources less than 2 miles beneath the surface.

There is much more to be developed, from 3 miles deep or more. The Massachusets Institute of Technology scientists concluded that heat mining methods could yield nearly 2000 times the power that the nation consumes each year. A developer would drill a well and use high pressure water to open fractures in the rock, which may be 300 degrees Fahrenheit. Then, injection wells would be drilled to circulate the water in the man made reservoir, and steam would be extracted to the surface to run electric turbines.

If a few technical problems could be conquered, power could be produced at from 3.6 to 9.2 cents per KWH. Alta Rock Energy in Seattle plans a demonstration project in the U.S. in the next 2 years. Their goal is to provide 10,000 megawatts within 10 years, enough power for 10 million homes and more than 4 times as much as old style geothermal plants produce.

If we could develop 5% of the geothermal power available to a depth of 2 miles, we could generate enough electricity for 260 million Americans. The government says this is possible by 2050.

Typically, hot water and steam is piped to the surface to drive a turbine and generate electricity. The temperature has to be at least 200 degrees F, the hotter the better. California, Nevada, Idaho, and Oregon have enough hot spots to supply most of their electrical power. Steam from the geysers is over 400 degrees F.

It is not easy to find a good source of hot water. It costs 2 to 3 million dollars to drill and you may end up with nothing. Doug Glaspey of US Geothermal estimates it costs 3.5 million dollars per megawatt to build a power station.

Geothermal plants have energy efficiencies of 8 to 15 per cent. It costs twice as much as it would in a coal plant, which produces power that sells for 5 cents a KWH.

Sites that are not hot enough to run turbines can be used to heat buildings with hot water. If you draw a line from North Dakota to Texas, almost all of the states west of the line have access to sources where the temperature is at least 200 degrees F. We are using only about 1% of the available sources adequate to heat our homes. A heat pump pumps the water into the home to supply heat and hot water. Potentially, almost all homes in the US could be heated this way. 85% of Iceland's homes are heated with geothermal heat.

The 2007 Energy Act authorized 95 million dollars for geothermal research. The cost of a heat pump for the average size house is $7,500. It will pay for itself in 10 years. There are roughly 1 million pumps installed in the US, and 50 to 60 thousand are being added every year.

Chapter 7

Solar Energy

Solar installations provide 0.039 % of the world's energy. Somebody suggested that the U.S. could supply its entire energy needs by covering a mere 1.6% of its land area with solar cells Photovoltaic cell power has grown by an average of 41% a year over the past 3 years Wind has increased about 18% a year. Photovoltaic cost estimates run from 23 cents to 32 cents per KWH, while residential electricity prices in this country range from 5.8 to 16.7 cents.

Pacific Gas and Electric announced a bold plan to install nearly 5 times the amount of solar power that is now operating across the country. They will use mirrors to concentrate some of the highest intensity sunlight in the world in the southwest U.S. The arrays will heat water to drive turbines just as in a regular power plant. With these slightly curved mirror troughs, power towers and photovoltaic arrays, they will produce 1.8 gigawatts of electricity. A gigawatt is 1 billion watts. Southern California Edison has contracted for a series of power towers that could provide 1.3 gigawatts.

Just outside of Seville, Spain, there is a field of mirrors, each one 1,290 sq. ft., that are used to direct the sun's rays to the top of a concrete tower, where the heat turns water into steam, which is used to drive turbines that generate electricity, 11 megawatts, enough to power 600 homes. The company hopes to increase that 10 times to serve all of Seville. There are 624 mirrors with this tower. A second tower with twice as many mirrors has been built next to it.

We need a bold plan to increase solar power in this country. It has been suggested that if 5% of the suns radiation on our Southwest desert areas, could be converted to electrical power, it would provide all we needed in 2006.

Eicke Weber, director of the Fraunhofer Institute for solar energy systems, in Freiburg, Germany, says,"The total power needs of humans on earth are approximately 16 terawatts per year. A terawatt is a trillion watts. By 2020, it is expected to grow to 20 terawatts.

Huge tracts of land would need to be covered with photo voltaic panels and solar heating troughs. A direct current "back bone" would need to be constructed to transport the electricity across the nation. By 2050, this could provide 69% of our electricity and 35% of its entire energy consumption, including transportation.

The power could be sold to consumers at 5 cents per KWH. Further development of wind and geotherrmal sources could take care of the rest of our need for electricity and 90% of the energy.

The federal government would have to invest 400 billion dollars to complete this plan over the next 40 years.

Solar plants consume little or no fuel, saving billions of dollars year after year. This would displace 300 large coal fired plants and 300 more large gas plants, and all the fuel they consume. It would no longer be necessary to import oil. In 2050, CO_2 emissions would be 62% lower than in 2005.

30,000 square miles of photo voltaic arrays would have to be erected. There is one disadvantage to solar. If it's cloudy or at night, no electricity will be made. Their plan is to use some of the electricity to compress air and pump it into underground storage. This would increase the cost 3 or 4 cents a KWH. The compressed air is released on demand to turn turbines aided by small amounts of natural gas. We already store our natural gas underground. This would be similar. The air would be pressurized at 1100 pounds per sq. in..

Another idea is to store heat produced during sunlight in molten salt and at night use this heat to provide steam for the turbine.

Near Las Vegas, there is a solar plant on 250 acres with gently curved mirrors which follow the sun and focus sunlight on oil containing long steel pipes which are heated as high as 750 degrees Fahrenheit. There are 182,000 mirrors. The heated oil flows into giant radiators that boil water into steam to drive a turbine and generates 640,000 megawatts of electricity to add to the grid—enough to electrify 14,000 households. A megawatt is 1 million watts.

The other way to harness sunlight is with photovoltaic panels, made of semi- conductors such as silicon, which convert sunlight directly into electricity. These can be placed on roof tops and the power generated can be used at the site or added to the grid. In Germany, an individual who has solar panels installed on his roof, will be paid for any excess power generated, and gets 8% of the cost of his installation for 20 years.

Also near Las Vegas and Nellis Air Force Base is a photovoltaic installation that supplies 25% of the electricity used at the base. It is the largest such installation in the US, but only the 25th largest in the world. There are 72,416 sun tracking panels. It generates 14.2 megawatts. There are larger plants in Germany and Spain.

Right now, solar is such a small portion of the power generated, .039%. It is expected to grow so that by 2030, it will be 10 to 20 per cent. Solar plants now receive a 30% federal tax credit to offset construction costs, and this will continue to be necessary.

CHAPTER 8

Wind power

No renewable energy is growing faster than wind power, but they only account for 1% of the country's electricity. To get to 20% as president Bush suggested, one new windmill would have to be built, every l5 minutes for 25 years, says GE's Vic Abate. Wind, solar and ethanol now provide 3.6% of the nation's energy. The government predicts that by 2030, it will be 4.2 %.

In The 1970s, wind turbine blades were 5 to 10 meters long, at the most. Each one would generate 200 to 300 KWH. It cost $2.00 per KWH. Now the blades are up to 40 meters long, [43 yards], and turbines produce up to 2.5 megawatts each, at a cost of 5 to 8 cents per KWH. The problem with wind energy is that sometimes the wind doesn't blow. Xcel Energy has plans to use large batteries, developed in Japan, using sodium and sulfur, and are as large as 2 semi trailers stacked. When the wind blows, batteries will charge, and current can be used at peak times of consumption or when the wind dies down. By 2020, Xcel expects to get 30% of its electricity from wind or about 3900 megawatts of power. One megawatt can supply as many as 1,000 homes. The batteries can last for 15 years.

Chapter 9

Fuel Efficient Vehicles

Congress just passed a law to require automobiles to average 35 mpg by 2020. The current standard is 27.5 mpg. The European Union set their goal at 44.2 mpg.

Japan intends to attain 45 mpg. These goals are expected to be achieved by producing more hybrid vehicles to make up for the larger gas guzzlers.

Natural gas is plentiful and clean, but expensive. Over 5 million vehicles around the world run on natural gas. You can put a small compressor appliance in your garage and fuel the automobile every night, from gas that is already piped to your house. Today, about 85% of Singapore's electricity is generated with natural gas. In the 1970s, an Exxon representative testified in Congress, that the U.S. had only 30 trillion cubic feet of natural gas remaining, and the fuel use act was passed in 1978, prohibiting the use of natural gas to generate electricity. Since then, we have produced 585 trillion cubic feet, and it is estimated that we have 1500 to 2000 trillion cu. ft. remaining.

World wide, car ownership is growing around 5% a year, so if emissions from cars are to be cut, engines will have to become much more efficient, or there will have to be replacement of gasoline by a cleaner fuel.

The Toyota Prius seems to be the state of the art so far in hybrid cars, with its battery and electric motor running much of the time and the gasoline engine running only when more power is needed. They may get up to 65 mpg. A unique feature is that the battery recharges when you step on the brakes as energy from the wheels is used to run a generator.

On the news it was announced that a prototype hybrid car has been developed, using a Saturn SUV with a battery that you can charge by plugging into a socket in your garage every night. It will go 80 mph on the electric motor and can go 40 miles before the battery runs down, and the gasoline engine takes over. It is expected to cost $8500 more than a regular SUV, and if you order it today it will be ready for you in 3 years. All you people living along the coast—take note. Do your part to save your homes.

There are other cars now available or planned for the future. The Honda Clarity gets 74 mpg and has no tailpipe emissions. Hydrogen, stored in a tank, mixes with oxygen in a fuel cell about the size of a carry-on suitcase. It produces electrical power that runs a motor, that spins the wheels. A Lithium battery kicks on from time to time to supplement the power. It drives like a regular car with the luxury features of an Acura. It can travel 280 miles on a tank of hydrogen at a cost per mile less than half of gasoline, It can be leased for $600 a month in a test program. Purchase costs are higher and it will require a chain of H2 filling stations.

The Super Prius from Toyota can be recharged at home and can go all electric for 10 miles. Then the battery will operate like the earlier Prius, switching between battery and gasoline engine as necessary. They also plan on an all electric commuter car with a range of about 50 miles, to be released in 2012.

BMW's car is BMW Hydrogen 7. The car has 2 fuel tanks, and 2 fuel gauges on the dash. There's a button on the steering wheel, labeled H2, by which you can change from hydrogen to gasoline or vice-versa. This does entail reduced power and limited fuel storage.

The Chevrolet Volt is an electric car that may be charged up at home. It has a larger lithium battery which will power the car for 40 miles. Then a small gasoline engine will kick-in for you to complete your trip. Charging the car will cost less than $1.00 a day. The price may be around $40,000.

The Nissan Leaf is an all electric car selling for around $24,000 to start. It can be plugged in at home to recharge.

The Ford Fusion Hybrid sells for around $25,300 and gets 47 mpg in electric mode and 41 mpg over all. It has a nickel hydride battery recharged while braking.

Toyota Camry and an SUV, the Highlander are available as hybrids. Chevrolet has an SUV, the Tahoe as a hybrid, priced at $37,280. Chevrolet Sprint gets 48mpg as a hybrid. The Honda Insight gets 53 mpg and Honda Civic as hybrid gets 46 mpg. Several trucks are also available as hybrid vehicles.

On the news, they announced a new car being produced in India, called the TAT car which is being sold for 2500 dollars. It is small and not luxurious, but is a closed car. No mention was made about mileage or what kind of motor it has.

390,000 Americans own a Toyota Prius. Toyota will have 3 cars in the Prius Line, a small car, a family car, and a utility vehicle. Chevrolet will come out with the Volt, a plug-in hybrid in 2010. Currently, it costs about $3000 more to go hybrid.

Taxicabs in New York City are slated to go green. By 2012, 13,000 taxis are to be hybrid vehicles. Ford Crown Victorias were getting 14 mpg. New vehicles are required to get 25 mpg, and the following year, 30 mpg. The plan will cut emissions in half in the next 10 years.

When it comes to electric cars, Have you heard about the Zenn? This stands for zero emissions, no noise. It is made in Canada, has a top speed of 25 mph, and can go 35 miles on one charge of its battery which can be charged from an electric outlet. It looks like a small car, with a convertible top. About 35 cities in Wisconsin have authorized its use on their streets where the speed limit is 35 mph. It costs $12,700. The battery can be 80% recharged in 4 hours, fully charged in 8 hours.

It can carry two passengers and a load of groceries. It has full weather protection, a heater, and air conditioner. It is powered by six 12 volt lead acid batteries. It costs about $200 to operate for one year. Its wheelbase is 81.8 inches.

In France, they have developed a car that runs on compressed air. It will be introduced in 2, 4, and 6 cylinder models, ranging in price from $4,800 to $12,900. One model, presumably the largest, fits 5 passengers, and can go 500 miles on the highway with one fill-up, as fast as 68 mph. Tank can be refilled in 4 hours. For more information, consult aircar.com.

The Tesla is an electric sports car that uses a lithium ion battery. It can go from 0 to 60 mph in 4 seconds, and costs $100,000. The first 350 cars were presold. Maximum range is 250 miles.

A Caltech scientist, Sosina Haile, has developed a new type of fuel cell, using a "super protonic" compound. The current prototype produces only enough energy to power a 100 watt bulb, but it may lead to a powerful fuel cell for automobiles. The chief emission is water.

Chapter 10

Recycling

For years we have been recycling to save energy and cut down on our carbon foot print. Is it important? An analysis in Britain showed it reduces CO_2 emissions by 10 to 15 metric tons a year, roughly equivalent to taking 35 million cars off the road.

Perhaps the most valuable benefit is the saving in energy and the reduction of green house gases and pollution when scrap materials are substituted for virgin feed stock. If you can use recycled materials, you don't have to mine as much, or cut trees or drill for as much oil.

Extracting metals from ores is extremely energy intensive. Recycling aluminum can reduce energy consumption 95%. It's about 70% for plastics, 60% for steel, 40% for paper, and 30% for glass.

Have you switched to compact fluorescent lights yet? They use 75% less energy. Pay back for spiral fluorescent bulbs is about one year.

CHAPTER 11

Acidification of the Ocean

A new development which also requires our attention, is that the oceans are becoming more acidic. As they absorb more CO_2 and other green house gases, they are gradually becoming more acidic. Some scientists fear the change may be irreversible. Sea creatures are at risk from phytoplankton to whales.

The oceans are already 30% more acidic than they were at the beginning of the industrial revolution. By the end of this century, they could be 150% more acidic. Measurements of Ph are taken from a buoy anchored in 5,000 ft. of water.

The concern is for the fisheries. Increasing acidity can eat away at the shells of crabs, oysters, and clams and put at risk the survival of nearly microscopic organisms known as "krill".It also inhibits calcification, the process by which these animals build their shells. Without shells, most of them probably would die. Squid also are sensitive to higher acidity, which affects their blood circulation and respiration. Colonies of coral would disappear.

Richard Feeley, an oceanographer with the National Oceanic and Atmospheric Administration in Seattle, said 500 million to 1 billion people, worldwide depend on fish for survival. Sharp declines in fish populations would severely affect their lives.

Eventually, acidification will reach inland waters, affecting oyster beds and clamming areas. Congress is acting on a bill to authorize research and monitoring of oceanic acidification. Scientists say, cuts in CO_2 emissions might slow global warming, but it could take thousands of years to reverse the increased acidity of the oceans.

Chapter 12

The Arctic and the Antarctic

I assume that we are all aware that there is no landmass at the North Pole as there is at the South pole, the continent of Antarctica. There is only the Arctic Ocean at the North pole. The Arctic sea ice by spring, gets about 10 ft thick and about 6 ft thick along the shores of the continents. During the summer, it begins to melt and break up. As the earth warms, the blocks of ice drift apart and it makes it difficult for the polar bears, who need to stand on the ice to hunt for seals. As more sea ice melts, their season for hunting seals is shortened, and they have to move inland and eat from garbage dumps.

In Canada's Hudson Bay area, it is estimated that there are now 25% fewer polar bears than there were 25 years ago. With the ice breaking up earlier, they have a shorter period on which to feed on seals and grow fat, so they can sustain pregnancies, and care for their young. The current population of 20 to 25 thousand could be gone by the end of this century.

There are landmasses in the far north, such as Greenland. It is an area about 3X as large as Texas. About 85% of it is covered with an ice cap, I mile thick. In some areas, it may be 2 miles thick. The ice melt there has increased 7% each year since 1996. Whether it will continue at this rate is questionable, But since the increase in climate warming CO_2 is higher than what was projected by the UN Panel on Climate Change, it is likely that it will..

The continent of Antarctica is buried under nearly 9000 ft. of ice in it's eastern portion, and 14000 ft. in West Antarctica. Two bays cut into Antarctica, the Ross Sea and the Weddell Sea. Both bays are filled with ice 600 to 1000 ft. thick. The outer half of the Ross Sea is nearly free of ice from December through February.

Much of the Antarctic Ocean freezes over in winter, forming a solid pack that stretches for hundreds of miles seaward. In October, the ice breaks up into chunks, and ice bergs break off the edge of the continent.

Penguins suffer from warming temperatures, because the rains soak the baby penguins, and they are chilled and at risk of freezing to death. Today, where there are 3000 pairs of Emperor penguins, by the end of the century, it is predicted that there will be only 400 pairs. In the past 50 years, the Adelie penguins, the ones with the black and white 'tuxedoes', have declined 65%. In the past 50 years, there has been a 6 degree rise in temperature in Western Antarctica, the largest increase in the world, causing more precipitation to fall as rain rather than snow. Penguin chicks, still in their down, get soaked through and die of hypothermia.

Ten ice sheets on the Antarctic Peninsula have receded or collapsed since the 1990s. The Wilkins Sheet is poised to break up, held in place by a sliver of ice 500 meters, [1640 ft.] wide, compared to 100 kilometers, [60 miles], in the 1950s.

Satellites see a double Texas sized loss in Arctic sea ice. Each year, during the month of September, the amount of ice floating in the Arctic Ocean is the lowest for the entire year. There is 20% less than was seen in the year 2000, or about 502,000 sq. miles less.

Chapter 13

A No Fossil Fuel Plan

In December, 2009, the leaders of the nations of the world met in Copenhagen, Denmark, to establish a plan to cut green-house gases. The best approach to this problem is to markedly reduce the consumption of fossil fuels and increase clean energy, with the use of wind, solar, and water systems, so that 100% of our electricity is produced that way within 10 years.

Nuclear power, because of waste storage difficulties, and the security risk, is not recommended in this plan. The plan was developed by Mark Z. Jacobson, professor of Civil and environmental engineering at Stanford University, and Mark A. Delucchi, research scientist at the Institute of Transportation Studies at the University of California, Davis.

The plan calls for millions of wind turbines, water machines, and solar installations. They have assumed that most fossil fuel heating stoves and ovens are replaced with electrical, and fossil fuel transportation can be replaced by battery and fuel cell vehicles.

The maximum amount of power consumed at any moment in the world is estimated to be 12.5 trillion watts, or terawatts. By 2030, the world will be using 16.9 TW, about 2.8 TW in the US If the world's power came only from wind, water, and solar, this amount would be less—about 11.5 TW, and US demand would be 1.8 TW. This is because in most cases, electricity is a more efficient use of energy. For example, only 17 to 20 percent of the energy in gasoline is used to move the vehicle, whereas, 75 to 86 percent of the electricity in an electric vehicle, goes into producing motion.

Now, we generate 0.02 TW of wind power, and 0.008 TW of solar energy, world-wide. This could expand to 40 to 85 TW for wind and 380 TW for solar. The plan calls for 9% of the demand, 1.1TW, to be met by water related methods—hydroelectric power, tidal turbines, and geothermal sources. Wind would supply 51%, 5.8 TW, and solar, 40%, 4.6 TW. Worldwide, here is the break-down of the number of plants that will be required in 2030, the size of the plants, and the percent that is already in place. They say that this would require less than 1% of the earth's land for the wind turbines and about 0.33 % for concentrated solar plants and photovoltaic other than rooftop installations.

490,00	Tidal Turbines	1 MW	Less than 1% in place
5,350	Geothermal Plants	100 MW	2%
900	Hydroelectric	1 MW	70% in place xx
3,800,000	Wind Turbines	5 MW	1% in place,
1,700,000,000	Rooftop Photovoltaic	0.003 MW	Less than 1% in place
49,000	Concentrated Solar	300 MW	Less than 1% in place
40,000	Photovoltaic Plants	300 MW	Less than 1% in place

The US share of this infrastructure would be:

76,685	Tidal Turbines
837	Geothermal Plants
141	Hydroelectric Plants
594,700	Wind Turbines
112,600	Wave Converters
266,050,000	Rooftop Photovoltaic Systems
7,668	Concentrated Solar Power Plants
6,260	Photovoltaic Power Plants

There is a possibility of scarcity of silver for solar cells and platinum for fuel cells, and other rare metals used in wind turbines and solar cells. They have looked into the cost of generating power in 2030 under this plan. Solar power would be the most expensive at 8 to 10 cents per KWH, but wind and wave and hydroelectric power should be about 4 cents per KWH.

Overall construction cost might be around 100 trillion dollars, worldwide, over 20 years. Most of this would be born by private investors who would make money selling electricity. It will take time to build all these plants and wind installations, but it took time to build our present system also. Remember, If we stick with fossil fuels, the demand would be 16.9 TW, requiring 13,000 large new coal plants which would occupy a lot more land, and what a disaster that would be.

What about "clean coal"—carbon capture and storage?

There are 3 steps.

1] capturing the CO_2 from a plant's smoke stack before it escapes.

2] compressing it into a liquid

3]transporting that through a pipeline, and storing the compressed CO_2 underground for hundreds or thousands of years.

It is very costly and controversial. It is estimated that it will cost from $15 billion to $30 billion to make it a reality. Most environmentalists would rather see the money spent on wind and solar projects. Additional coal would have to be burned in the capture and storage steps.

President Obama's $787 billion stimulus package includes $70 billion for renewable forms of energy, research, and weatherization of federal buildings and homes. The administration supports the construction of a Smart Grid so that electricity, generated in remote areas of the desert can be transported to populated areas. It will cost $5400 billion over 10 years, and is projected to save $46,117 billion in 20 years.

Chapter 14

Methane

As a result of global warming, the permafrost in Siberia and Canada is melting. As it melts, it forms lakes and ponds, and as the animal and plant matter in it begins to rot, due to bacterial and fungal action, it releases methane, which bubbles to the surface. Methane, CH4, is 25 times more powerful at heating the planet than CO_2, molecule for molecule. and could greatly enhance the rate of global warming.

According to Vladimir Romanovsky of Fairbanks, Alaska, permafrost temperature has been rising since the 1970s. He calculates that one third to one half of the permafrost in Alaska is within one to one and a half degrees Celsius of thawing.

Katey Walter Anthony, research professor at the University of Alaska, Fairbanks Water and Environmental Research Center, and her group, have been studying the problem for about nine years. She believes the release of methane is a significant factor in global warming. Vladimir Alexeef, also from Fairbanks, has said that according to his best estimates it will raise the temperature of the earth an additional 0.32 degrees C. Permafrost covers about 20% of the earth.

In Siberia, the "skin" of the earth, made up of muddy, mossy peat, is described as being only half a meter thick, [19 and a half inches] and under it, the frozen ground extends 40 to 80 meters deep. As the thawing progresses, sink holes develop, and are filled with water runoff. Northern Siberia is about 30% lakes. An accumulation of wooly rhinoceros, mammoth, lions, bears, and horses and the grass they fed upon, are in the permafrost.

Katey Walter Anthony was able to tell from the white bubbles in the ice, where the seeps at the bottom of the lake were. As she describes it, "I stabbed an iron spear into one white pocket, and a wind rushed upward. I struck a match, which ignited a flame, five meters high, knocking me backward, burning my face, and singeing my eyebrows. Methane!!!" She has collected and measured the amount of methane from seeps on 60 lakes, and has caught as much as 8 gallons a day from some seeps. She has mapped these lakes with their seeps and kept records through the years of her study. The radio carbon age of the gas, is 433,000 years old. Over the course of her study, the amount of methane has been accelerating.

Slowing the emissions of CO_2 is the only way to slow the process of melting permafrost. These researchers predict that if CO_2 emissions continue to increase at the current rate, northern lakes will release 100 million to 200 million tons of methane a year by 2100, much more than 14 million to 35 million tons they emit annually today.

Chapter 15

Making Coal Cleaner

We have enough coal to last for a hundred years in this country. But to utilize it poses a serious problem of contributing to global warming. About 80% of CO_2 emissions produced by making electricity, comes from burning coal. Also there is the problem of soot which falls on ice and snow and causes it to melt faster. Some states have banned new coal plants, and many companies have canceled their plans to build coal powered plants.

The greatest hope for continuing coal consumption has been carbon capture and storage or CCS. There are no plants yet which do not emit CO_2. They need to capture the CO_2 as it is released from the stack, compress it into a liquid, and transport it to underground storage.

At a coal fired power plant in New Haven. W. VA, It is hoped by the end of 2009, it will be an example of CCS. Before, it emitted 8.5 million metric tons of CO_2 annually. They have drilled deep wells for storage, and they add special chemicals to the smoke stack to separate CO_2 from the other emissions. Initially, they hope to capture and store 100,000 to 300,000 metric tons of CO_2 annually, and go up from there. This is 1% to 3% of the plant's emissions.

It has taken years of research and false starts, exciting break-throughs, and thousands of hours of lab testing and analysis. Sites for storage have to be secure to keep CO_2 in place for hundreds or thousands of years. Salty aquifers are ideal because they tend not to leak. It is difficult to find a proper storage site.

In Australia, there were plans to build a 2 billion dollar coal fired plant that captured 90% of it's CO_2 emissions, but it was abandoned when they found that the storage site had too many cracks.

A project to build such a plant in Illinois, known as Future Gen, was abandoned when President Bush pulled some 1.3 billion dollars because of cost over-runs. Environmentalists would rather see the money spent on wind and solar energy plants.

Chapter 16

Burning Garbage to Produce Syngas

In the process of plasma gasification, torches pass an electric current through a gas, even ordinary air in a chamber, to create a superheated plasma—an ionized gas with a temperature of 7000 degrees Celsius, hotter than the surface of the sun. This is lightning in a bottle, which can break down chemical bonds of garbage placed inside the chamber, converting organic compounds into syngas, {a combination of carbon monoxide and hydrogen} which can be used as fuel in a turbine to generate electricity. It can also be used to create ethanol, methanol and biodiesel. The slag, resulting from the process, can be made into construction materials.

Major plants, capable of processing 1000 tons or more, daily, are under development in Florida, Louisiana, and California. Syngas generated electricity has a smaller carbon footprint than coal. For every ton of trash processed, you reduce the CO_2 emissions by 2 tons. If all the garbage were put through this process, we could produce 5% to 8% of our total electrical needs—equivalent to 25 nuclear plants, or all of our current hydroelectric power output, says Louis Circeo, director of Plasma Research at Georgia Tech. Research Institute. With the US expected to generate a million tons of garbage every day by 2020, using plasma to reclaim some of that energy could be too important to pass up.

Chapter 17

Other Effects of Global Warming

Already in Bangladesh, a man tasting the water, says it tastes salty, Five years ago,it was good sweet water. Water from the Bay of Bengal is surging up the fresh water rivers and percolating into the surrounding soil, spoiling it for rice production. Floods have been heavier than normal and have killed fifteen hundred people and damaged about two million tons of food. The United Nations warns that a quarter of Bangladesh coastline could be covered with water if the sea rises three feet in the next fifty years.

Some success has been attained by developing a salt tolerant rice variety. They would like some remuneration from the high polluting countries. At the meeting in Copenhagen the rich high emitting countries agreed to provide $30 billion in the next 3 years to help poor countries cope with climate change and a $100 billion a year by 2020. They also agreed to cut their emissions at least 80% by 2050.

As a result of global warming and warm currents in the ocean, giant jellyfish, weighing as much 400 pounds, have swam northward, and are ruining the fishing industry. Fishermen bring up their nets, crowded with blobs of jelly, and they have to lift these huge blobs of jelly and throw them overboard, avoiding their poisonous tentacles, which have already stung and spoiled some of the fish. Many fishermen in Japan's Wakasa Bay have just given up and quit fishing. They are also a problem in the Middle East and Africa.

In 2007, a salmon farm in Northern Ireland lost more than 100,000 fish to an attack by the "mauve stinger", a jellyfish normally known for stinging bathers in warm Mediterranean waters. Scientists say they have migrated to Irish seas because of global warming.

In Japan, invasions of jellyfish have cost the industry approximately 332 million dollars a year. The waters of the Yellow Sea have warmed as much as 1.7 degrees Celsius over the past quarter century. Agricultural and sewage run-off insure a strong growth of plankton, which the jellyfish feed on.

Oceans absorb about 25 per cent of the green house gases pumped into the atmosphere, from human activities, according to a new report from the UN. That helps slow global warming, but it forms carbonic acid, further aggravating the effects of ocean acidification. Ocean acidity could increase 150 per cent by 2050.

As we have milder winters, there is less oxygen in the deeper parts of the ocean. This will affect marine life including the fish. Some 1 billion people in the world depend on fish as the main source of protein. The problem of ocean acidification is as important as climate change.

If we continue to burn fossil fuels at the same rate, the ocean will become more acidic than it has been for 65 million years. Scientists have studied the ocean bottom and found shells of marine organisms and established this fact. If it happens again, there will be a mass extinction of marine life.

In the past, there has been warming of the earth at times gradually over millions of years. The rate at which the green house gases and warming are increasing now, is 100 times as fast as happened naturally in the past. Humans are doing in centuries what natural processes can do in millions of years.

As global warming develops, the agricultural production in the tropics will decrease, but it will increase in the northern parts of the world. By the latter third of the 21st century, the food supply will have to be raised another two times, to feed all the people who are expected to be alive then.

Economists say, to have an energy system that does not emit CO_2, would cost 2% of our wealth each year. Now let's suppose we have such a system. If you ask yourself, "Is it worth melting the ice caps and acidifying the oceans so you can be 2% richer?" I think the answer is obvious.

This must not be a political issue, even though a Democrat, Al Gore, first called it to our attention, Republicans and Democrats must unite to save our planet.

A question was asked, "What is the most compelling evidence that human behavior is actually warming the earth?" Professor Kenneth Caldera of Stanford University answered,."It is the fact that the stratosphere, the upper atmosphere, is cooling while the lower atmosphere and the land surface are warming. This is a sign that greenhouse gases are trapping energy and keeping that energy close to the surface of the earth."

One of the most serious effects of climate change is the melting of the glaciers in the Himalaya Mountains. For centuries, these glaciers and the melt therefrom, has fed the greatest rivers in Asia, the Yangtze, the Yellow, Mekong, and the Ganges. On Mt. Kowagebo, 22,113 ft. high, there has been a glacier that has been retreating for many years, but its pace has accelerated in the past ten years, over a football field per year.

Something like 2 billion people in 12 countries, are dependent on the melted snow and ice from these mountains for their water supply, from the arid plains of Pakistan to the metropolises of China, 3000 miles away. The Tibetan Plateau is heating up twice as fast as the global average of 1.3 degrees F over the past century.

The ice cover has shrunk 6% on the Tibetan plateau since the 1970s, and in Tajikistan and northern India, 35% and 25%, respectively. In the western part of Tibet, the glaciers are still growing a little. Chinese scientists fear that 40% of their glaciers could be gone by 2050, and it will lead to an ecological catastrophe.

On the northern part of the Tibetan plateau, people are already affected by a warmer climate. The grasslands and the wetlands are deteriorating and the permafrost has receded to higher elevations. Thousands of lakes have dried up. Desert now covers one sixth of the plateau. Herders, who once thrived there, cannot find pasture for their stock. In Qinghai province, tens of thousands of nomads have had to give up their way of life,and move into substandard housing, and get by on a small government dole.

By contrast, on the southern edge of the plateau, there is an abundance of water from the increasing melt, and expanded croplands and longer growing seasons, and production has increased. As the melt continues, eventually there will be a sharp drop in run off, with an acute water shortage, and a plunge in production. China plans to build 59 reservoirs to capture glacial run off in an area just north of the Tibetan plateau.

The Maldives, located in the Indian Ocean, around 1200 tropical islands, with snow white beaches, swaying palm trees, and richly colored coral reefs, stretch across more than 600 miles. Eighty per cent of these islands are no more than 3.3 ft. above sea level. These will all be covered with water if the oceans rise as much as climate scientists predict.

According to most recent reports, whereas it had been previously predicted that sea levels might rise 1 to 2 ft. by the end of the century, they now estimate 3 to 6 ft. as the ice is melting faster than expected. Some studies warn that the Southwest US will become a dust bowl after 2040, and Kansas will be above 90 degrees for 120 days of the year.

CHAPTER 18

The Problem of Nitrous Oxide

Another green house gas that must be considered is nitrous oxide, which, by weight, is 300 times worse in it's effect on global warming than CO_2. It' s a by product of fertilizer and other industrial processes. Much of the fertilizer used on the fields is washed off into the streams and rivers and ends up in the ocean, where it causes bloom of the algae, that utilizes the oxygen, causing depletion of it's availability for the fish and other sea creatures. causing a dead zone at the mouth of the Mississippi river and other rivers all over the world.

Like chlorofluorocarbons, [CFCs] it is a threat to the ozone layer also. We have solved the CFC problem by no longer using them in spray cans. Currently there is no attempt to reduce nitrous oxide, because it is not regulated by the Montreal Protocol, which regulates pollutants. Each year, scientists have estimated that nearly 1 billion metric tons of CO_2 equivalent are released globally.

Besides the millions of tons of fertilizer, nitrous oxide comes from livestock manure, sewage treatment, and automobiles. Actually, two thirds of nitrous oxide emissions come naturally from the earth itself, as bacteria in soil and the oceans break down nitrogen.

What can be done about this problem? Reducing the quantity of fertilizer used in farming, switching to a less meat heavy diet, and lowering the number of cars on the road, while boosting fuel economy, will all help.

Chapter 19

Update and Overview

The economic stimulus package provided $21.5 billion for scientific energy research. We need laws which would set a price on carbon, or a cap and trade program, and one that would require that we get a certain percentage of our electricity from renewable sources, such as wind and solar. This would encourage investors to put money into these projects.

The government has ear-marked $70 billion in the stimulus bill for green energy and for work on buildings to make them more energy efficient, upgrading public transit, and building an electric grid, on which to transport electricity from remote areas to where it is needed. Researchers from The Political Economy Research Institute estimate that $100 billion in green investment would create 2 million jobs.

Young people are being trained in our colleges to be able to install solar panels on the roofs of houses. The classes are filled to capacity. Expect a big increase in solar installations on roofs.

First Solar, the largest solar panel maker in the US, is building a massive solar field in China. E-solar, an American company, has contracted to assist in building a concentrated solar mirrors and tower installation in China, that will generate 2,000 megawatts. China plans to reduce CO_2 emissions 40 to 45 per cent from 2005 levels by 2020.

Currently, they get 9% of the power from wind, solar, and hydropower. They expect to increase this to 15% by 2020. Coal provides 2/3 of China's power and is expected to be the dominant source for some time.

Climate scientist, Jon Foley, of the University of Minnesota, calls for an

80% reduction in CO_2 by 2050. Removing CO_2 directly from the air would help to achieve this goal.

Buildings account for 2/5 of the energy we use, and account for 2/5 of the CO_2 emitted in this country. Better insulation, use of natural light and ventilation and a ground source heat pump are ways of reducing energy use.

The ground source heat pump is based on the fact that the ground stays between 40 and 50 degrees, down about 15 feet, so if you put tubing down there, you can use the warmer air to warm your house in the winter and the cooler air to cool your house in the summer. It works on the same principle as your refrigerator.

Producing power with wind and solar requires storage of electricity. The development of the smart grid will make it possible to transport power to populated areas where the power is needed. Every time a light is turned on or off, the generation plant has to wiggle up or down. Now, when the generators are producing too much, the markets have to pay somebody to take the extra energy. The price ranges from $20 to $500 per megawatt hour. The grid will level this out. Sometimes in Texas, wind farms have to shut off in the middle of the night, because they can't get the electricity to the station. There is not enough demand where the wind farms are, and there is nowhere to put it.

According to T. Boone Pickens, we are currently getting 85 million barrels of oil a day world-wide. The US uses 20% of all the oil used, with only 5% of the world's population We are importing 60%..People are trying to cut down on consumption. In 2005, 200,000 hybrid cars were sold in the US.

We are about half way through all the oil in the world. We have produced about a trillion barrels and there's about a trillion more. It will be pretty well gone by 2100, Use of natural gas is growing 25 to 30 per cent a year. He says its 30% cheaper than gasoline. We have a good supply of it in this country. He says it burns 86% cleaner, but he may be thinking of particulate matter. Actually, a mole of natural gas or methane, CH4, when burned, produces 1.12 moles of CO_2. Since CO_2 is about 3 times as heavy as methane, one pound of natural gas would produce about 3 and a third pounds of CO_2 when burned.

Pickens says that Los Angeles has about 2500 buses and 2100 of them run on natural gas. He proposes that all semi trucks now in service, continue to burn Diesel, but that all new tractors built should use natural gas.

It is expensive to convert a car to natural gas. The cost runs from $12,500 to $22,500. Honda produces its Civic as a natural gas car that costs $25,190, whereas a regular Honda Civic is almost $10,000 cheaper. There would have to be more natural gas filling stations throughout the country to make its use feasible. Twenty years from now we would be importing natural gas. Today about 20% of our power is from plants that burn natural gas. It is the most expensive power produced. 50% of the power generated is from coal.

If we have a target of generating 25% of electricity by renewables by 2025, we could push solar to 4% and wind to 10%. Fossil fuels would still be producing 75% of our electricity.

Earth has been warmed by 0.75 degrees Celsius, [1.35 degrees F] over the past century. If we doubled the CO_2 in parts per million from 280 to 560, it would raise the average temperature, 3 degrees. European politicians have agreed that CO_2 should not rise above 450 ppm and temperature not more than 2 degrees Celsius more than pre-industrial levels. We're at 387 ppm right now at the end of 2009. CO_2 is increasing 2 ppm per year.

Physicist, Myles Allen, of the University of Oxford, says, to keep warming below 2 degrees C, humanity can afford to put one trillion tons into the atmosphere by 2050. It has already put half of that in. Only one quarter of known deposits of coal and oil and natural gas can be burned. Emissions need to fall 2 to 2.5 per cent per year from now on. Japan's prime minister says his nation will try for a level of emissions 25% less than 1990 levels by 2020.

Brazil now ranks fourth in the world in carbon emissions, and most of it comes from deforestation. Demand for biofuels is causing a dramatic expansion of Brazilian agriculture, which is invading the Amazon at an alarming rate. The price of soybeans goes up; the forest comes down. Deforestation accounts for 20% of all current carbon emissions in the world. This was an important subject for discussion at the Copenhagen conference and the rich nations were trying to appropriate money to help save the rain forests of the world. Biofuels do reduce our dependence on foreign oil, but the basic problem is that using land to grow fuel, leads to the destruction of forest, wet lands and grassland that store enormous amounts of carbon.

When sunlight hits a leaf, it splits water into hydrogen and oxygen, through the process of photosynthesis. The hydrogen is stored as fuel. For years scientists have been trying to understand this process. Daniel Nocera, a chemist at M.I.T., has discovered a catalyst, cobalt phosphate, which can drive the process of photosynthesis outside the leaf, in the lab, and does it cheaply.

To quote him, "To take care of the average house in a day, you need 20 kilowatt hours of electricity, which is equivalent to only 5.5 liters of water. To drive that point home. I'm holding the amount of water in my hands that you need to power a very big house on the California coast. That amount of water takes care of that house as well as powering a fuel cell car around town. So that's the future. There's no way to stop it. Nature already did this 2 billion year experiment and decided on this process, and it's coming soon."

Chapter 20

What Can We Do About It?

No nuclear plants have been built for 30 years. In 2005, the Bush administration ear marked $30 billion for nuclear development. Recently, President Obama committed $8.3 billion of this money, in loan guarantees for the construction of two new nuclear reactors in Georgia. There are more than two dozen US projects awaiting approval by the Nuclear Regulatory Commission. The Dept. of Energy is requesting $36 billion more. A reactor costs about ten billion dollars. Obama says one nuclear plant will cut CO_2 emissions by 16 million tons each year, when compared to a similar coal plant. That's like taking 3.5 million cars off the road.

Another aspect of nuclear power are the miniature nuclear reactors developed by engineers at Oregon State University, and are to be produced by Nu Scale Power. They are small enough to be mass- produced, and can be shipped by truck or boat. They can be sealed like a big battery and put underground for as much as 30 years. Then they can be removed and taken to a storage facility or a waste processing plant.

They may put out as little as a few megawatts instead of the 1,000 megawatts of conventional plants. Hundreds of similar devices are already operating around the globe. Some are laboratory test reactors; others power submarines, ships, and even US and Russian military outposts.

The reactor is pressurized and filled with water that flows past the core where the radioactive decay of uranium 235 generates intense heat. The heat boils a separate tank of water, and steam is produced to drive a turbine and produce electricity.

Through stimulus spending, the Obama administration has begun to combat global warming with 6.1 billion dollars appropriated for developing advanced hybrid vehicles, 10.5 billion for grid modernization, 18.1 billion for mass transit, 19.9 billion towards energy efficiency and 26.6 billion to renewable energy generation.

Finland plans a massive energy boost. Three utilities have made requests to build new nuclear plants, but they will have to be approved by parliament. The European Union requires that all nations must raise the share of renewables to 38% of energy by 2020. Finland plans to increase wood burning, wind power, bio-fuel production, and the use of heat pumps to heat buildings. Their efforts are expected to reduce CO_2 emissions by seven million tons by 2020. Reducing the use of coal will cause a further decrease in CO_2 emissions of two million tons.

China is spending 9 billion dollars a month on clean energy projects. China's coal related CO_2 emissions are projected to reach 9.3 billion metric tons or 52% of the world's total by 2030. India's emissions will reach 7 % and the US, 14%, according to the Energy Information Administration. In the US, the wind and solar industries had record years in 2009. In 2008, the United States surpassed Germany as the top producer of wind turbines. In 2009, China became the top producer of wind turbines.

General Electric is the top manufacturer in the US. But they have had to cut back, because the market for them has slowed. T Boone Pickens has given up, for the present,of proceeding with his planned large wind farm in Texas, because there is no way he can send the power to the large cities across the state.

The energy bill is in Congress now and it would set up a cap and trade plan to attempt to control CO_2 emissions. In the absence of a federal plan for expansion, the states have set up their own plans. Thirty states have passed laws, requiring a certain percentage of power to be from renewables.

Recent polls have shown that the percentage of people who believe that global warming is caused by human activity, has fallen from 47% to 36%. The Tea Party movement is partially responsible for this as they tend to be hostile to global warming science. The IPCC does know what its science shows and there is a consensus of most of the scientists that the earth is getting warmer and it is due to human activity, particularly, an increase in green houses gases, especially, CO_2 from burning fossil fuels.

Here are some odd facts that are interesting and pertinent. When a car uses 1 gallon of gasoline 19.4 pounds of CO_2 is produced. Diesel fuel would produce 22.2 pounds of CO_2. For every mole of natural gas that is burned, 1.12 moles of CO_2 are produced. Burning 1 million BTU of natural gas will release about 117 pounds of CO_2. When I see a train go by with over a hundred cars I always think that that is a good way to transport stuff. 43% of America's freight moves between cities by rail. One train can carry the same load as 280 semi trucks. Freight trains can move a ton of cargo 457 miles on a gallon of fuel. A tractor trailer can do it 130 miles on one gallon. It takes 1200 to 1800 BTU to transport a passenger one mile by high-.speed rail. Conventional trains consume 2600 BTU per passenger mile. Air travel is 3,300 per passenger mile, and cars are 3,500 BTU.

Soot is also a factor in global warming. There is much emphasis on the effect of CO_2 on climate change, but another substance, soot, may be almost as dangerous. The fine black powdery pollutant comes from Diesel engines, power plants, the clearing of forests and burning of fields, and open cook stoves which are common in developing countries. When it falls on the snow and ice, it absorbs sunlight and hastens the melting. In the past the heating effect of soot was mollified by the effects of sulfate particles, generated by coal and petroleum combustion. The light colored sulfates strongly reflect sunlight, cooling the ground. However, measures to combat acid rain, due to the sulfates, have largely eliminated this cooling effect. Soot's warming effect is felt mainly in the Northern Hemisphere, because Antarctica is so far from populated centers. Studies have indicated that soot is as big a contributor to melting of the Himalaya glaciers as CO_2. The Arctic could have some year around navigable shipping lanes by 2030. Soot from increasing shipping traffic could further darken what snow remains. However, capturing soot is a lot easier than controlling CO_2. Auto-makers are already starting to use particle traps that cut tail-pipe emissions from Diesel engines by 90%. So called "bag house" filters can capture soot from coal fired power plants.

In China and India, they are switching from coal stoves to nearly soot free gas burning cookers. CO_2 remains in the atmosphere for hundreds of years, but soot can be scrubbed out by rain or snow-fall. If enough bag house filters are used, in a few months the air will be clear of soot. In May, 2010, the eight nation Arctic Council, which includes the United States, recommended clamping down on soot. Congress is considering legislation to reduce it. To check climate change, controlling black carbon may be a relatively simple and some what effective fix.

CHAPTER 21

Removing CO_2 From the Atmosphere

CO_2 can be removed from the atmosphere by a simple chemical reaction with sodium hydroxide, NaOH, better known as lye. It is not available in stores, but can be bought on the internet. It is caustic, and if a child or an animal swallows it, it causes scarring and narrowing of the throat passages. For that reason, one must be careful in its use.

For more than a year, I have been hanging a small blanket, 24 inches by 64 inches, soaked in sodium hydroxide solution, from a line in my back yard. It becomes dry in several hours and I assume that the chemical reaction ends then. I resoak the blanket in the solution and hang it again, so it is done twice a day, morning and evening. This procedure takes only a few minutes each time.

I use plastic gloves to protect my hands. If you are careful in putting them on and removing them a glove will last 3 or 4 weeks. Get the thumb and fingers headed in the right direction and pull them on by grasping the cuff. When you take them off, pull on the fingers and thumb and pull them straight off. They do not get very wet, but you will get some of the solution on your fingers when you remove them. Just go and rinse off your hands under the faucet immediately. I have never noticed any burning but I always rinse as a precaution. If you notice that your hand is wet while inside the glove, discard the glove for it has a hole in it somewhere.

I use a 5 gallon pail to hold the solution. I use a gallon milk jug to add the sodium hydroxide, 1/2 level teaspoon per gallon. I catch a couple of inches of water in the jug, put in 1 level teaspoon of the NAOH, put the cap back on and shake it to dissolve and mix it, then fill the jug and pour it into the pail.

I try to avoid "gurgling" to minimize splashing. Then I fill the jug with water and pour that into the bucket. That makes as heavy a load as you will want to carry. Use a fully extended arm to carry it comfortably.

I use a couple of clothes pins to hang the blanket or it could be a large towel. I dip it in the bucket and poke it with my fingers a few times, hopefully to get some of the carbonate out of the blanket . Then I grab a corner with the right hand and make a circle with the thumb and fingers of the left hand and I strip it until I come to the more bulbous part and I squeeze this with my hand and continue stripping on down to the last 6 or 8 inches. which I squeeze. No wringing is necessary. I loop it back into the right hand as I proceed. I hang it and it drips a little but I haven't noticed any effect on the lawn. Do not let the dog accompany you as it may be tempted to catch the dripping.

After you have done this a few times, there will not be much solution left in the pail. Go ahead and mix another 2 gallon batch. After you have used maybe 3 or 4 batches, you will see a fair amount of precipitate in the bottom of the bucket. You can pour this into a wide mouth bottle or a quart jar and let it sit. The carbonate will settle to the bottom. You can decant the clear liquid carefully without losing the carbonate before you pour in the next batch.

What if it rains? This may just renew the chemical reaction with any residual NaOH in the blanket. A heavy rain will probably cause it to drip and you will lose both NaOH and $Na_2 CO_3$. The rule is that if it is already wet you don't need to do your chore.

What about cold weather and freezing? Then you have to bring your act in doors. You will need a room that you can keep the kids and the pets out of. You will need to string a line across the room or you can make a wood frame with braces on the upper corners and string a line across it. You will need to buy a window planter box with plugs in the drain holes, about 8 inches wide and long enough to accommodate the blanket or towel you are hanging. The solution can drip into the box.

You must keep the bucket in a safe place where no pets or children can get into it. I set mine on another 5 gallon pail that has a lid on, to keep it up off the floor of the garage. In winter, you can leave it in the secure room.

I drive my Chevrolet Malibu 13 miles to work every day and 13 miles back home again. Driving 55mph I get about 26 miles to a gallon of gas. Burning one gallon of gasoline produces 19 pounds of CO_2. I guess I am a big polluter. Think of how big a bubble of CO_2 this must be that I release into the atmosphere each day. It is a gas and how much of it would you need to weigh 19 pounds. A lot of you drive farther than that every day.

I thought maybe I could reduce my carbon footprint by dipping a blanket, but I find that I can only remove a few milligrams this way. It would help if this was done in all of the back yards all over the US and China and all over the world. It shows how important it is to do all of these other things. Unfortunately, I am afraid that they will not be done. It is almost impossible to get a meaningful energy bill passed in our congress. They want to delay any cap and trade agreement because it may retard our recovery from the recession, and I am inclined to agree that it could be delayed for a year or two. However building the solar plants and wind farms and nuclear plants will provide jobs and help the recovery.

Some scientists are working on machines that will remove CO_2 but the minute you hook up a fan to blow air through, you are using energy and adding more to your carbon footprint. If you want to do something personally, using your own energy, to help save the polar bears and the penguins, dipping and hanging a blanket is one thing you can do. Storage of sodium carbonate is not much of a problem because the volume that you get is small. People speak of storing CO_2 underground. The gas requires so much space. It isn't practical. In the form of sodium carbonate it could be stored easily. It is used to manufacture sodium bicarbonate. It is used as a water softener, and it is used in making tooth powder and tooth paste. It is some what alkaline but should be pretty safe to deal with. You could sprinkle it on your garden or lawn, or dig a hole and bury it. Please don't try to use it for tooth powder.

The author is an 85 year old retired physician who specialized in internal medicine. Because of his habit of reading medical literature, when he retired from medical practice, he became interested in the problem of global warming, and read everything he could find on the subject.

It appeared to him that it was a serious problem, and he was concerned that in the past two years, polls showed that the percentage of people in the US who believed that it was due to human activity, had gone from 46% to 37%. He wanted to write a book that everyone could understand, so that they might come to believe, and see that we do what we can to prevent it.

He was born and grew up on a Wisconsin dairy farm, with 7 brothers and sisters. After Pearl Harbor 2 of his 3 older brothers became naval officers, and the other one, a marine officer. When the marine captain came hone to the farm, Elton went into the army, and served a year in the Japanese occupation. After service, he started pre medical classes in Platteville State Teachers College and transferred to the University of Wisconsin in Madison to finish. He attended medical school and graduated with an MD in 1954. He practiced medicine, mainly in Costa Mesa, California for 30 years.

He has a sister, Mary Eastwood who, with Betty Friedan, founded NOW and started the movement for women's rights. An Attorney working for the Dept. of Justice, In Washington, DC, in her off duty hours, she wrote the briefs for some of the earliest cases for equal pay for women.

His younger brother also went to Washington, DC and was the head of the Dairy Extension division of the USDA for many years. A younger sister married a Lutheran minister and they have unleashed 8 more highly educated Meyers on the world.

Elton married and they had 3 children. After 15 years, they were divorced. He was married again to a girl who grew up in Duluth, Minnesota, and always wanted to go back there. In 1989, He and his new family of 2 adopted daughters, moved to Duluth. After he had a knee replacement, he worked at temporary jobs as a physician in several cities in Wisconsin, Michigan, and Iowa.

His wife developed breast cancer and had surgery and radiation therapy. She went 5 years without trouble, but then it recurred and she died two years later in 2003. In 2000, they had moved to Stone Lake, Wisconsin where Elton still lives and takes care of their boxer dog and two cats.

He loves to play golf and hopes to break a hundred this year. He plays 9 holes about 3 times a week, and usually walks, He also works out every morning before going to work as a Realtor in Hayward. He loves to watch the Packers and Badgers football games and the Pro Golf tour on the TV.